30DAYS
TO MORE PATIENTS *and*
A MORE ORGANIZED OFFICE

By Teresa Korenstein

Copyright 2011

Liability Waiver: The information in this guide does not represent the only method to succeeding in a private practice. Individuals using this manual waive any claim they may have against Terri Consulting Group or Teresa Korenstein for any problems or damage that may result, in any way, from their reliance on the information presented in this guide.

Introduction

Congratulations! You've taken your first steps towards having the successful, profitable, and reputable practice you've always wanted. This program was created for you, the private practice owner who cares about his patients and wants to see his practice grow. During the next 30 days, you will be given assignments that market and organize your practice into the successful healthcare business you know it can be. I'll give you real world examples throughout this guide that support the efficacy of these tasks. If you are diligent and put in the effort, you will achieve great results.

I recommend that you begin this program on a Monday and do the program from Monday through Friday each week. If you want to speed up the process and work through the weekends, that's fine, but most people want that time to rest and recharge. You may have purchased this book alone, or as part of a package from Terri Consulting Group, LLC (TCG). In that case, TCG is with you every step of the way. Contact TCG as much as you need to during this process. You are never limited to your scheduled consultation meetings. If you have purchased this book alone, you can always add on consulting services by contacting TCG at info@HealthMarketingStrategies.com. You can find out more information at www.HealthMarketingStrategies.com. However, rest assured that you will still receive a great benefit by purchasing this book alone, and a great boost to your business.

This program assists you in marketing and organizing your office. Regarding marketing, we're not just going to market to Orthopedic surgeons. There are so many Physical Therapy referral sources out there that you don't just have to buy doctors lunches in order to get patients. Let's explore the following diagram:

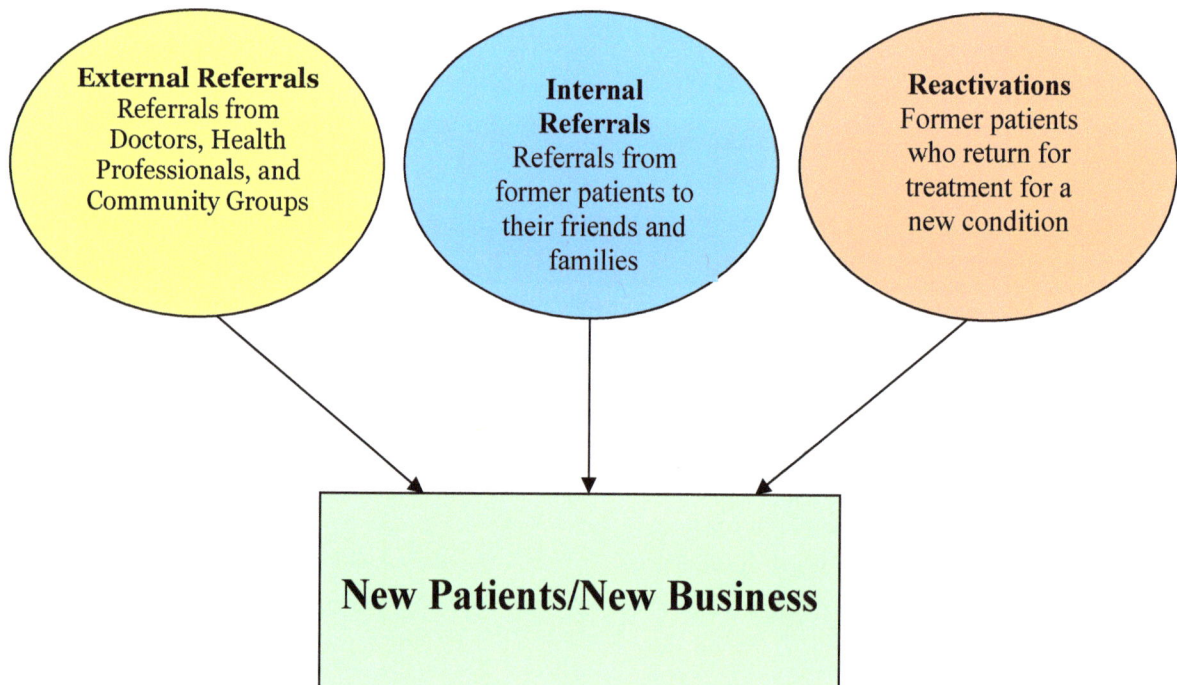

External Referrals
Referrals from Doctors, Health Professionals, and Community Groups

Internal Referrals
Referrals from former patients to their friends and families

Reactivations
Former patients who return for treatment for a new condition

New Patients/New Business

External Referrals, Internal Referrals, and Reactivations all bring in New Patients to your clinic. New patients are new business and that's what we all want. The more new patients you have, the greater the demand for your services. Your office will be busier, and your potential for profit increases.

External referrals are referrals from anyone outside of your business. These include referrals from Doctors, Health Professionals, community leaders, and community groups. Internal referrals are referrals that are generated in house. Mostly these are referrals from former patients who recommend you to their friends and family. They may also come from your staff. The third group is Reactivations. A former patient who is discharged from Physical Therapy but decides to return for a new condition is considered a Reactivation.

You don't have to spend a lot of money marketing to these three groups. The key to effectively marketing your Physical Therapy practice is to watch your trends and reinforce what works for your clinic. Every office is different. What works for you may not work for someone else. That's why TCG includes extensive consultations throughout your 30 days. Your consultations are where you personalize your marketing plans to the specific needs of your office.

This program is not only about marketing, it's also about getting your office organized for success. During the next 30 days we will spend time training staff and setting up office operations. This is important because we can get any number of patients to want an appointment in your office, but, if you don't know how to schedule them and hold onto them, your office will never prosper. Consider the following:

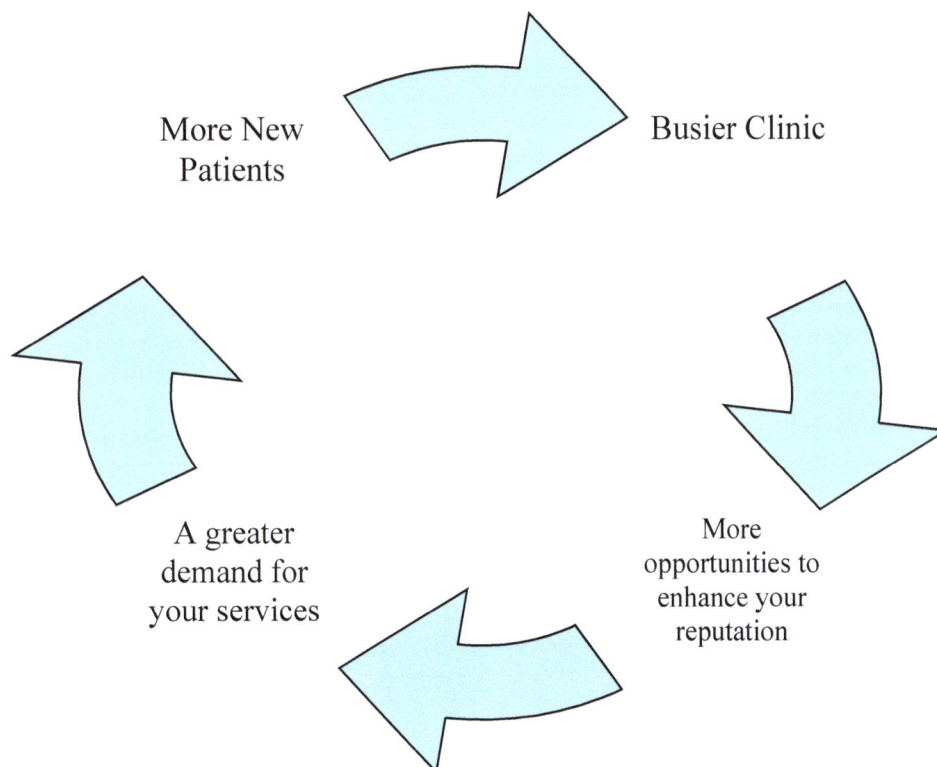

More New Patients → Busier Clinic

Busier Clinic → More opportunities to enhance your reputation

More opportunities to enhance your reputation → A greater demand for your services

A greater demand for your services → More New Patients

If your marketing efforts result in more new patients, you will have a busier clinic. Each new patient brings an opportunity to enhance your reputation by impressing the patient, his referring physician, and his friends and family. The greater your reputation, the greater the demand for your services. However, if you and your staff can't get those patients scheduled and keep them on the schedule, you will lose business, opportunities, and the potential for growth. Bringing in business is only half the challenge. You also need to know how to keep the business.

The calendar on the next page shows each day's tasks at a glance. Each day's assignments are designed to be completed during one business day, unless otherwise noted. You may find it a little hectic at times, but I encourage you to put as much effort as you can into these next few weeks.

So, get ready because in 30 business days, you may no longer recognize your current business. In its place you may find the more organized, more successful, and more profitable practice of your dreams. Without hesitation then, let's begin! Take a peek at what's in store for Day 1 and contact your TCG representative to begin planning meetings. Good luck!

30 Days To More Patients and a More Organized Office

Sunday	Monday	Tuesday	Wednesday	Thursday	Friday	Saturday
	Day 1 Staff meeting Audit office aesthetics.	**Day 2** Go through charts.	**Day 3** Choose and train a PL.	**Day 4** Customize Marketing Survey.	**Day 5** Check next week's schedule. Create New Patient Packet	
	Day 6 PL coaching and calling current patients. Go through charts	**Day 7** Write out 30 Thank You notes.	**Day 8** Create and schedule Class.	**Day 9** Network with municipal workers. Distribute flyers.	**Day 10** Create Trend Report. Check next week's schedule.	
	Day 11 Execute Marketing Plan. Go through charts.	**Day 12** Begin sending Thank You postcards or surveys from patients to Doctors.	**Day 13** Train PL to call patients after 1-2 weeks of treatment.	**Day 14** Develop Step Down Program.	**Day 15** Create Trend Report. Check schedule. Create Newsletter.	
	Day 16 Execute Marketing Plan. Go through charts.	**Day 17** Audit website and create Facebook page.	**Day 18** Set up Referral Rewards Program.	**Day 19** Determine how you will utilize the next holiday in your marketing plan.	**Day 20** Create Freebie coupon. Create Trend Report. Check schedule.	
	Day 21 Execute Marketing Plan. Go through charts.	**Day 22** Develop system for tracking Discharge Calls.	**Day 23** PL to make follow up calls. Give out Freebie coupons.	**Day 24** Give out Freebie coupons. Execute Referral Rewards Program.	**Day 25** Write out Job Descriptions. Create Trend Report. Check schedule.	
	Day 26 Execute Marketing Plan. Go through charts.	**Day 27** Establish Perfect Attendance Rewards.	**Day 28** Write out birthday cards for upcoming month.	**Day 29** Make calls to discharged patients.	**Day 30** Celebrate your patients (and your hard work) with Patient Appreciation Day.	

Day 1

This is an exciting day! Today you lay the groundwork for your future success. Your office is on the verge of becoming very busy, and you have to be sure it's ready for growth. Today we talk to your staff and get your entire team on board for this plan, *30 Days to More Patients and a More Organized Office*. A lot of people resist change, especially when they feel it's being forced upon them. If you invite your coworkers to participate in the process, they may be more accepting. Therefore, your first assignment is to schedule a staff meeting for today. You should use the Staff Meeting Guide in Appendix 1 to direct this meeting.

Take it very seriously if you have a staff member who is really complaining about this plan to you or to others. Schedule a meeting with this employee immediately and hear their concerns with an open mind. If they don't come around, you may have to let them go. You need your team's support for this program, as well as any other program you hope to execute.

Throughout this guide I will explain the purpose of each assignment and support it with real life experiences. I once worked with an office that had seemingly reached a plateau despite having a very popular, very skilled clinician as Director. No matter what marketing strategies they tried, the office didn't grow. After a few months of this we began looking at different possibilities. Well, they could have spent a fortune trying every marketing idea out there and nothing would have made that office grow. It turned out that the very popular PT Director did not want growth. She didn't like it if the office was too busy. She discharged as steadily as she acquired new patients so her volume never changed. We were able to talk about this and get her assistance so she didn't discharge prematurely. We also proactively put another clinician on the schedule. As soon as this happened, the volume grew and this clinic really flourished. Both schedules filled up. Since I've been there, they've also added a third clinician part time. So this is really something you have to talk about with your team.

At the end of the staff meeting, invite everyone to participate in an exercise. The purpose of the exercise is to get everyone's feedback on the aesthetics in your office. Once again, involving everyone is good for morale.

Today's exercise is to get some feedback on the visual appeal of your office. Sometimes when you spend so much time in a place, you can't see its flaws, or its potential. This kind of thing really matters to patients. I remember the experience of having a patient come in for her Initial Evaluation, and not make any more appointments. We called to follow up and she said, "I have to be honest with you. The staff is great but the clinic looks terrible. If you can't take care of your office, how can you take care of me?" We apologized but the damage was done. She never returned. Not everyone will give you such an honest critique. Most people will just ignore your calls and give their business to someone else. She did us a huge favor by giving us feedback. We immediately went to work getting rid of clutter and setting up a cleaning and maintenance schedule.

For your meeting, let everyone know their honest feedback is valued and appreciated. Distribute the sheets in Appendix 2 and use them to critique the office.

You should also enlist a trusted friend to do the same activity and report back. Your friend should answer the questions on Appendix 3.

At the end of this exercise, share everyone's worksheets. What things can you do to make the place cleaner, neater, and more efficient? Develop a checklist of cleanup and organization tasks that need to be done regularly. Assign someone for each task and set the parameters for frequency. Appendix 4 may be useful.

Day 1 Checklist:
- ☐ Have a staff meeting to get everyone on board for your new program.
- ☐ Run an audit of the aesthetics of your office and gym.
- ☐ Have your co-workers and an objective friend go through the office and report findings. Develop a plan to maintain the standards you've discussed.
- ☐ Create a task list to maintain the appearance if your practice.

Day 2

Today's task is to go through every active chart and account for the patient's status. Most of you do this already but we need to do it today so you have a clean slate moving forward from this point. If you have a computer program that runs this for you, that's great. I'm a little old fashioned and like leafing through the paper charts. Any method is fine, as long as you accomplish the goal of knowing the status of every patient. Are they scheduled? Should they be scheduled? Are they discharged or on hold? If you don't know what's happened to a patient, call them and note on their file that you've called. If a patient is discharged, update their status. If their status is hold, put a note on the calendar to remind you to call them at a later date.

The purpose of this task is twofold. First it allows you to keep track of patients. It's embarrassing and unprofessional to have another healthcare provider contact you about a patient and you don't know what happened to them. A doctor once called an office I was working with and asked about a patient he had referred. What had the patient been doing in PT? Not only had there been no recent Progress Note on this patient, but there was no record of them coming in for the past 3 weeks. The PT hadn't called the patient, and there were no updates in the file. They simply lost track of the patient. They had to confess that they hadn't seen the patient in 3 weeks and didn't know how the patient was doing. Needless to say the doctor was not pleased and referrals from his office declined. It would have been so much better if they could at least say they left messages for the patient.

When you're doing this task, keep an eye on your Worker's Compensation patients. Case Managers handling Workers' Compensation claims need to know when their patients cancel appointments. Case Managers are a great referral source and Workers Compensation plans usually reimburse well. Don't overlook this avenue of potential referrals and keep track of your patients. Maintain good communication with Case Managers and they will remember that you are easy to work with and communicate well. We'll talk more about this in the upcoming weeks.

The second reason that this task is important is because it allows you to do damage control. A very busy clinic I worked with frequently lost track of patients. When they started doing this task, they heard from patients who were dissatisfied with their progress and had issues with the staff. Obviously, you don't want to hear negative feedback, but it is better for you to hear it and address it than having the patient tell their friends, neighbors, and physician.

I also want to touch on what to do when you hear negative feedback. Very often the person complaining just wants to be heard. After they have said their piece, it can be quite effective to say, "Thank you for this feedback. I'm sorry you had such a frustrating experience. I hope you will allow me to correct the matter." Some patients won't come back after they feel they have been wronged, but at least they will appreciate your polite response. Nothing is as detrimental to your reputation, nor as unproductive, as arguing.

Day 2 Checklist:
- ☐ Go through patient charts and account for the status of each patient.
- ☐ Establish who will be responsible for doing this task each week and when they will do it.

Day 3

A phone call is very often the first interface a patient has with your facility. Therefore, your office's phone etiquette has to be perfect. So many offices lose visits because patients don't get through on the phone lines, have to wait too long for an appointment, or they're just shopping around and like the voice of your competitor's receptionist better than yours. I remember working with an office and hearing a patient tell me that the reason they chose the office was because the receptionist sounded sympathetic on the phone. It can really be that subjective! You need to put someone in charge of the phones because this is important to your business. We're going to call this person you choose, a Patient Liaison (PL). This may be the receptionist or some other person on your team who has time to answer phones. This person should have excellent phone etiquette, be extremely friendly, and a great listener.

Today you're going to spend time sitting with this person and training them. They should be able to answer any question they're asked about you and your practice. Go through the questions on Appendix 5 with them, and add your own. The PL has to know the ins and outs of what you do. The PL should never sound hurried or rushed. Your PL shouldn't have to put patients on hold to get answers to basic questions. Most importantly, your PL should not be someone who makes comments like, "We've been so busy lately!" and, "It's a zoo in here!" "Zoo" is not a favorable impression for patients. Even though you probably give great care in your "zoo" environment, it's not something you want said about your business to patients who might repeat these words to their physicians. Imagine a patient going back to a doctor and saying, "I'm doing great at Best Physical Therapy, but boy it's a madhouse there! They're so busy!" What image do you think stands out, that the patient said they're doing great or that he described the office as a madhouse? These comments could discourage a doctor from sending more patients. Afterall, the doctor wants to know that his surgery is being protected and his patient is getting the best possible outcome. Since he's not present at your clinic, he relies on feedback from patients and the results he sees. For this reason, your staff needs to be careful how they describe the practice in front of patients. Some of you might be rolling your eyes because you already have a good receptionist. Well, good isn't enough. Your PL has to be amazing. This is the first impression of your business and nothing beats the first impression.

If the PL isn't the first line to the phone, they should be given the phone when a new patient calls in. Make sure everyone on your staff knows this. When a new patient calls, the PL should be handed the phone. Of course, it goes without saying that everyone on staff has to answer the phones on the first ring and with an enthusiastic greeting:

"Good morning! You've reached Best Physical Therapy, this is Terri. How may I help you today?"

When the patient says they want an appointment, immediately take their name and phone number so that if anything happens to the call, you can call them back. Cell phones drop a lot of calls and that could be lost business for you. Instruct the PL to gather the information that you need in friendly tones. If you're smiling, patients hear that smile through the phone.

Tell the entire staff to put the PL on the phone even when people are shopping. Healthcare is something we shop for now, just like everything else. Your PL needs to be your best sales person. Anyone who calls up "shopping" should be encouraged to talk, and come in for a tour.

Instruct the PL to get patients in as soon as possible. If a patient is in pain and you can't see them until the end of the week, they may try to go elsewhere. This is a lost opportunity. What's worse is that patient may go back to their doctor and say, "I tried to go to Best Physical Therapy, like you said, but, they were too busy. I'm going to Awesome Physical Therapy instead." This creates an opportunity for your competitor! Doctors don't want to hear that you can't see the patient they just referred, or that their patient has to wait. Tell the PL to come to you if it doesn't seem like the patient can get an appointment in the next 3 days. Try to make appropriate scheduling and staffing decisions so you can get the patient a timely appointment.

Throughout this program, we're going to assign the PL other tasks to do and oversee. They can be found in the Appendix 6 under PL's Weekly Schedule. The PL's main responsibility is maintaining relationships with current patients. Although the PL doesn't have to personally do every task, she should be sure the task is done completely and up to your standards.

Day 3 Checklist:
- ☐ Elect and train your PL.
- ☐ Make sure everyone on your staff knows and understands this position.

QUESTION: Why are we investing so much time maintaining our relationships with current patients?

ANSWER: Because each current patient has the potential to bring in new business.

Example: Mr. Smith is a current patient. The graphic below illustrates how he can bring you new business.

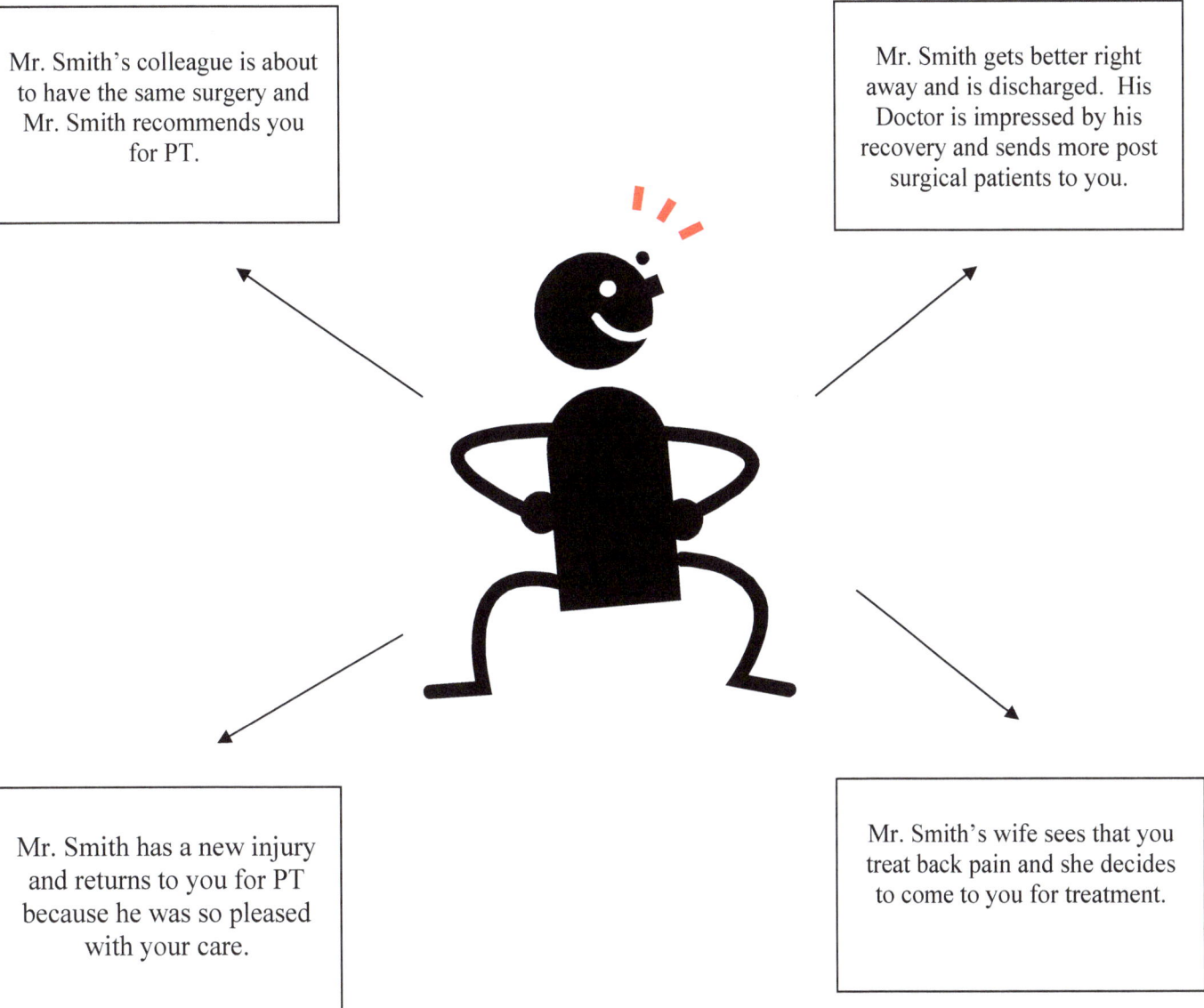

Mr. Smith's colleague is about to have the same surgery and Mr. Smith recommends you for PT.

Mr. Smith gets better right away and is discharged. His Doctor is impressed by his recovery and sends more post surgical patients to you.

Mr. Smith has a new injury and returns to you for PT because he was so pleased with your care.

Mr. Smith's wife sees that you treat back pain and she decides to come to you for treatment.

In this example, Mr. Smith has the potential to bring new business from his physician, his coworker, and his family, and he has the potential to be a Reactivation in the future.

Day 4

Have you heard the saying, you need to know where you've been to know where you're going? This is absolutely true in the PT world. Where are your referrals coming from? If you know this, you know where to reinvest your time and money.

There are so many potential referral sources out there. TCG breaks it down into three main groups – External Referrals, Internal Referrals, and Reactivations. External referrals can come in a lot of different forms, for example, community groups, local businesses, friends, neighbors, nurses, and of course, physicians. Most Physical Therapists focus only on the latter group. They spend their money, time, and energy marketing only to physicians. Unfortunately, you miss out on a lot of information this way. Consider the following, Dr. Klein may tell all his patients to go to your competitor despite your best marketing efforts. One day, a patient comes in with a prescription from Dr. Klein. Great! You feel like you are finally getting through to him. However, the real reason this patient came to you has nothing to do with Dr. Klein. The patient came to you because you treated his Personal Trainer and the Personal Trainer recommended you. Although this may not be what you want to hear, it's still an opportunity. You can talk to the Personal Trainer about setting up a two way referral system. Maybe you'll get more patients from him than you ever would have from Dr. Klein!

If you dig a little deeper into each referral, you'll learn about the strengths of your business and you'll create marketing opportunities. Let's say a patient walks in with a prescription from Dr. Jones. Dr. Jones has never referred to you before. You're thrilled with this opportunity, but before you get excited, find out some information. Did Dr. Jones tell the patient to go to your office? Your patient tells you that Dr. Jones has a list of PT practices he gives out and you are not on the list. However, none of the PT practices on the list accepted your patient's insurance so he found your office on his own. This is a golden opportunity! Now, when you market to Dr. Jones, you bring something very valuable to the table. I suggest visiting Dr. Jones's office immediately with a copy of your patient's Initial Evaluation note and tell Dr. Jones how this patient found you. He may start directly referring all of his patients with that insurance to you. And, hopefully, you'll get your name on his referral list and more of his patients will come your way. You can only get this kind of information if you dig a little deeper. If you simply saw the prescription as a straight referral, you would never have know about the list. Spend time being accurate at the front end, and you won't waste time later.

Today's task, then, is about collecting information. From today forward, all of your patients should fill out a Marketing Survey on their initial evaluation. There's a sample in Appendix 7. Customize this for your office and collect one from each patient. We will be consolidating the information next week and using the data to generate statistics.

Day 4 Checklist:
- ☐ Customize a Marketing Survey and give to each patient.

Day 5

If you began The 30 Day Program on Monday, then today is Friday. You are almost done with your first full work week! Today's task is lengthy and may need to be continued on the weekend. As I've mentioned, the 30 Day Program is designed to get you more patients from every avenue, not just Doctors. In order to get more new referrals, others need to know what you do. So today's task is to create a New Patient Packet. The purpose of the packet is to let new patients know all the ways you can help them, their friends, family, neighbors, and coworkers.

The New Patient Packet can be a folder, a tote bag, or just a collection of papers. You can also email this material if the patient prefers it. It's recommended that you include a Welcome Letter that sets the tone of your practice. This letter should be written from the perspective of the owner of the practice and include her or his contact info. Giving that information really shows that you care about your patients and want to hear their concerns. You should also list all the services you provide and conditions you treat. Please note that in the example in Appendix 8, the conditions are listed in medical and non-medical language such as "back pain" and "herniated disc pain." This is helpful so that potential patients can easily find their condition. Finally, your New Patient Packets ask for referrals in a discreet way. Appendix 8 shows 3 pages that should be in your Packet. These pages should be printed on your letterhead. Feel free to add any other materials such as practice brochures, flyers, and business cards.

The second task today is to look back on your schedule for the week and make sure that everyone who came in for the week is either scheduled for the following week or accounted for in some way. If anyone isn't scheduled who should be, call them and get them back on program. This job should be done every Friday. It can be delegated to an Aide or Receptionist.

Day 5 Checklist:
- ☐ Create a Welcome Packet for New Patients
- ☐ Review the past week's schedule and make sure patients are scheduled

Week 1 Review

Congratulations! You've made it through your first week. Let's review what you've accomplished:

- You've had a staff meeting to motivate your staff.
- You've done an audit of the aesthetics of your practice and made appropriate changes.
- You've decided who will maintain basic needs of the practice to keep it looking good.
- You updated the status of all of your current patients.
- You chose and trained a PL.
- You started keeping good records of your referral sources.
- You reviewed this week's schedule and compared it to next week's.
- You created a New Patient Packet that promotes your office.

Well done! You're on your way to more patients and more organization.

Day 6

Happy Monday! Let's jump right into your second full week. Remember we're using a 3 prong approach. We're increasing external referrals, strengthening internal referrals, and going after Reactivations. We're also reinforcing the importance of Physical Therapy to get the maximum participation from your current patients. If you get your office running well, and go after these avenues of business, your practice will grow.

Your current patients can be amazing referral sources. They have firsthand knowledge of your expertise. There are three points you need to make to new patients:

1. **You treat a variety of conditions.** You want your patients to find their friends and family in your service list. For example, they might notice that you treat balance problems and can help their Mother-in-law who fell recently. The best way to get this point across is to go through your New Patient Packet. Unfortunately, it's not enough to simply hand out this material or send the email. You need to explain it to new patients. An office I worked with stopped giving out New Patient Packets because they said they always found them in the waiting room. Nobody took it home and the office felt like it was a waste of time to produce them. They were right. It's a waste if you don't explain the purpose of the materials. There's so much paperwork in healthcare. Whatever you're giving out needs to be meaningful. Help your patients make that connection.

2. **Go through their prescription with them**. Most patients never read what's on the Rx. The most important thing you're pointing out is the recommended frequency and duration of Physical Therapy. You want your patients to come for every visit. Let the patient know what the doctor is recommending for optimal care. Think of this as good business sense. If you're supposed to get 12 visits from a patient, try to collect all 12 unless there are appropriate reasons for discharge. In my experience as a clinician, very few patients were in such great shape that they needed fewer visits than the original prescription. Usually, we're battling with insurance companies to get them more!

3. **Explain your cancellation policy.** If you have a cancellation policy, explain it on the first day. If you don't, remind patients how important it is to keep appointments and ask for the courtesy of a phone call if they can't keep an appointment.

The next task is for the Physical Therapist to make some phone calls. The Physical Therapist should call the patient in the evening after their Initial Evaluation, and express the following:

- "Thank you for coming to Best Physical Therapy. We're going to spend this week working on (protecting the injury, restoring motion, reducing pain) and the following weeks working on (building strength and mobility). You may be a little sore after your evaluation. I'm looking forward to seeing you (next appointment date and time). Please let me know at any time if your expectations are not being met. If you have any questions please hold onto them and we can discuss them when we're together. Have a good night."

This script works for nearly every injury. The purpose is to hold onto Evaluations. An Initial Evaluation is an investment. It requires extra work and time. If you don't already, you need to start thinking of your time as money. You book extra time for an Evaluation and you use a lot of one time resources. For example, there's time spent scheduling the patient, registering the patient, checking insurance benefits, writing notes, and performing the evaluation. These one time tasks pay off when you have the patient for long term treatment. Therefore, a patient that comes in just for the Evaluation and then never comes in again wastes time, and money. A phone call after the evaluation is a nice touch and helps manage the expectations of the patient. They may confess that they don't feel PT will help because they feel worse than they did before you saw them. You know what to say in these situations to keep the patient coming in.

I typically hear some groans about this task. If the patient tries to keep you on the phone for an extended time, just repeat that you will happily address their concerns in person.

The other phone call that the Physical Therapist needs to make after an Evaluation is to the Case Manager. If a patient has a Workers Compensation or No Fault claim, a Case Manager will typically be assigned to the case. As I've mentioned earlier, Case Managers are excellent sources of referral. Use every opportunity to call Case Managers. After the Evaluation, call the Case Manager and give them a brief 1-2 minute summary of the Evaluation and Plan of Care. Of course you will be faxing or mailing this information to them as well. You can also ask the Case Manager if there is a Job Demands Analysis (JDA) on file. This will tell you the demands of the patient's job so you can best tailor your treatment plan. The Case Manager will definitely appreciate the communication and your interest in getting the best outcome for the patient.

Since today is Monday, you should also review your charts. Run your report, or leaf through the paper charts and make sure you know the status of your patients. Follow up on patients who are on hold. We will be doing this every Monday. Since it was already done so thoroughly last week, it shouldn't take as long to do today.

Day 6 Checklist:
- ☐ Train your staff to:
  - ○ go through New Patient Packets with new patients
  - ○ tell new patients the recommended frequency and duration on their prescription
  - ○ explain your cancellation policy
- ☐ Start calling New Patients after their Evaluation to manage expectations and promote continued attendance
- ☐ Start calling Case Managers after a patient's Evaluation.

Day 7

Today we start going after some new business. Go back and look at your referral sources over the past 6 months. Put them in order from most referrals to least. This may take some time. Some software can generate this list for you. You want this to be as accurate as possible.

Next, get your favorite stationary or letterhead and write a note to the top 15 and the bottom 15 referral sources. Use Appendix 9 for examples of appropriate letters. You don't have to write a lot. Simply convey the message of, "Thank you." A lot of people ask if they can just send an email. A traditional letter shows that you put time and effort into the communication. If you always communicate with a particular doctor via email, that's fine, but not recommended.

We're going to hand deliver these thank you notes. If you have an Evaluation, Treatment, or Progress Note, for a specific doctor, hand deliver that, too. The purpose of this is to say thank you to the doctors that refer regularly and to keep you in the minds of both the ones that do, and the ones that don't. Your marketing needs to be consistent and effective. You have to stay in front of your referral sources without being obnoxious. No one wants their time wasted. Delivering a note is a valid reason to stop by an office and get you in front of referral sources.

Some offices may bring the doctor out to see you. In that case, don't give your sales pitch. Remember the purpose of your visit is just to say thank you for the support. Doctors are so used to being hit up for business by Drug and Product representatives. (If you don't believe me, ask any private practice nurse!) It's exhausting to be constantly solicited. You stand out because your visit has a different agenda. Keep your visit short, because patients may be waiting. Some good things to keep in mind when speaking to potential referral sources:

- **Be interested rather than interesting.** Doctors and their staff spend all day listening to others. If you spend some time listening to them, they'll remember it.

- **Make connections.** What are you more likely to remember next month, the funny conversation you had about your kids, or the statistics on the efficacy of a specific manual therapy technique? Most of us would remember the funny story. Making a connection with a referral source is invaluable. A Physical Therapist I worked with was excellent at that. He grew a huge multi-million dollar business with several satellite offices and he never once uttered a sales pitch. He simply talked to everyone with respect – nurses, receptionists, doctors, and patients - and made great connections about family and friends. He talked about sports, his children, their schools, and the community. He put people at ease and was well received. His results are proof enough that this works!

- **Don't forget the support staff.** Sometimes a powerful nurse is the real head of an office. Don't bulldoze over the support staff in an effort to get to the physician. The nurse or receptionist might be the one holding the cards. There are a few Orthopedists I know who don't make recommendations for Physical Therapy, but their nurses sure do.

Day 7 Checklist:

- ☐ Write Thank You notes to your top 15 and bottom 15 referral sources.
- ☐ Hand deliver the Thank You notes along with patient notes
- ☐ Make connections and be interested rather than interesting

Day 8

Today we may be going outside of your comfort zone. We've talked about external referral sources coming from places other than physicians, such as your community. You need to be out in the community and let everyone know the great things you do. One of the best ways to do this is by hosting a free information session on a popular topic. Today's task is to develop a 30 minute information session on a topic of your choice. Choose something that you are comfortable speaking about, and remember to try to appeal to as large a percentage of the population as possible. If the topic you choose to speak about is very specialized, you may not have a strong attendance. The talk can be very basic since the audience is not health professionals. When you are ready, put the class on the schedule, and create flyers to advertise it. Flyers do not need to be fancy. Most word processing programs provide templates for flyers. TCG is also happy to assist you with this.

One of the best topics that appeals to people of all ages is an information session on back pain. You know the statistics of how many people suffer with back pain. If you have a good relationship with a Chiropractor or Spine Specialist, invite them to participate and be a guest speaker. Maybe they can give the basic content and you can show the exercises you recommend or ways to protect your spine during everyday tasks such as putting groceries away, lifting, and sitting at a desk.

Day 8 Checklist
- ☐ Create a class for your community.
- ☐ Create flyers to advertise the class.

Building Your Reputation in the Community: Speaking Topics

- Osteoporosis Exercise Class
- Preventing Injury in Student Athletes
- Ergonomics of Working at the Computer
- Pregnancy Related Back Pain
- Shoulder Pain
- Runner's Knee Treatment and Prevention
- Tennis Elbow: How to Get Back in the Game
- Spine Clinic: Neck and Back Pain
- Maintaining Good Joint Health
- Golf Clinic: Improve your game and lose your pain

Day 9

Today we're going to advertise your class. Run off as many copies of your flyer as you can because today you're going to hit the road. The first stop is Town Hall. If you live in a town that has a Town Hall, it's imperative that you get to know the key players who work inside. I'm not recommending an entry in the political arena! Rather, it's a good idea to know them because they are typically popular, well connected, and have opinions that get heard and shared. It would be nice for you to have the mayor of your town as a patient. Maybe he or she will get the Police, Fire, and Public Works departments to come to you for Physical Therapy. So, take your flyer and go to Town Hall. Present your flyer to the first person you see and tell them about your class. Ask them if they would like a free class on this or some other topic presented at the town hall for employees of the town. Some places call this a "Lunch and Learn" seminar, if it is given at lunchtime. At this point you'll probably get referred to someone in the Recreation department or some other department. Follow the string of leads until you find the appropriate person to talk to about this. Don't let this idea drop. If you succeed and get the opportunity to present a class to the employees of the municipality, your reputation can grow quickly.

I did a lot of classes on Back Pain Treatment and Prevention with a clinic I worked with and it always produced a patient or two. Whenever we needed a boost in business, we scheduled this class. There are so many positive effects of the class. First, and most obvious, you get potential patients – people with back pain who want care. Second, you increase foot traffic to your clinic. People get to see your office and be impressed with your facility and services. Naturally, you're going to plug your practice and programs at the class. People will keep this in mind. I recommend giving everyone who attends a New Patient Packet. Someone might not come to you for their back pain, but they may hurt their knee in a month and decide to come to you for help. Third, you build your reputation as a Health Expert in your community. Even those who don't attend your class, will keep in mind that your practice gives lectures on back pain.

Over the next two days, you should visit the following places and ask to display your flyers:
- Orthopedists' offices
- Neurologists' offices
- General Practitioners
- Senior Groups in town
- Women's Groups in town
- Libraries
- Community Centers
- Community Boards at Churches and Synagogues
- Local hair salons

If you have a Patient's Evaluation or Progress Note, bring it along to hand deliver to a doctor's office, along with your flyers. You may need to enlist the help of your staff to get your flyers to all of these groups. Don't give up and don't be overwhelmed! Every time you talk about your clinic or your class, you help drum up business.

Please note that the flyers have to ask people to call and reserve a seat at the class. You must have names and phone numbers of everyone who wants to attend. Attendees should be called and reminded of the class 1-2 days prior to it. This helps turnout. You also need their phone numbers in case you have to reschedule the class for any reason.

If you have the means, you can consider Direct Mail to advertise the class. Sending out a flyer or postcard to residents can be effective but this is a more costly means of advertising. If you have any interest in this, TCG can help. Contact us for prices and referral information.

At some point on days 8 and 9, you also need to call your local paper and find out how to submit a press release about your class. Nothing beats free press! Some local papers are itching for stories for their health section. It's a great idea to get comfortable with the process of sending in press releases. There are usually a lot of samples online to help you with this.

Days 9 Checklist
- ☐ Go to the Town Hall, try to set up a Free Lunch and Learn or information session about back pain.
- ☐ Distribute your flyer to local doctors and in your community
- ☐ Call your local paper and send out a press release about your class.

Day 10

Remember those Marketing Surveys you have been so diligently collecting? Today you make them work for you. Use Appendix 10 to create a Trend Report from the data in the Marketing Surveys. When it's completed, think about the following:

- Are your patients coming mostly from doctors or from other patients?
- Is there a particular doctor who has been referring more than others?
- Are you surprised to see a doctor who referred?
- Were you expecting to see more referrals from a particular source?

Having an objective 3rd party look at this information can be extremely helpful. Those meeting with TCG should discuss this material during consultations. From your Trend Report, develop a Marketing Plan. Your Marketing Plan will include what stops you want to make to advertise your practice. For example, which doctor's offices do you want to go to. Do you want to setup a meeting or a lunch. Do you plan to visit any community groups or organizations. Consider the following:

- Whenever you stop by a doctor's office, you should bring a copy of the patient's most recent note. Medical notes are a great way to introduce yourself and provide a reason for your visit. If you're visiting a new referral source, you should also bring a copy of your New Patient Packet and a list of the insurances you participate with.
- Visit new referral sources to introduce yourself.
- Visit referral sources that have picked up and given you a lot of recent referrals. Be sure to say thanks.
- Visit referral sources that have dropped off. Make the purpose of your visit to give them updated insurance lists or some other information. You may need to setup a meeting or lunch with these types of referral sources to find out if something happened that caused their business to drop.

Because it is Friday, you should also be comparing this past week's schedule against next week's schedule. Are there patients who are missing? Call them to get them back on program.

Day 10 Checklist
- ☐ Create a Trend Report
- ☐ Create a Marketing Plan based on trend report
- ☐ Review last week's schedule and compare it against the upcoming week's schedule. Account for patients who are not scheduled.

Week 2 in Review

This week we went into the community and started tracking referrals. You achieved the following:

- Introduced a New Patient Packet that advertises your services and policies.
- Scheduled a class.
- Distributed flyers about your class to local professionals.
- Setup a class for Municipal workers.
- Created a Trend Report based on your Marketing Surveys.
- Delivered 30 Thank You notes to Doctors' offices.

In addition your office has changed procedures:

- Physical Therapists are calling new patients and Case Managers after each Initial Evaluation.
- New Patients are hearing about your cancellation policies.
- You are analyzing statistics each week.

Congratulation on making these changes! Change is never easy. Is anyone on your team complaining? The next few weeks get easier as you begin to repeat tasks. but don't lose your initial enthusiasm for this program. If you put in the time, energy, and effort, it will work!

Day 11

Here we are at Week 3. You're in the thick of it now. Don't give up! A more successful, more organized office is within your reach. You should be having regular staff meetings, and checking in with your staff to make sure they are on board with the changes you're making. It's Monday so have someone go through all your active patient files and update the status of each patient. Any active patient who isn't scheduled and isn't on hold, should be called.

Today we're going to implement the TCG marketing plan you developed last week. I am often asked how I feel about giving out goodies such as donuts, bagels, or candy to doctor's offices. I don't have a problem with any of that. Handing things out can make you memorable. Of course, I'm speaking only of small items such as pens with your logo, sticky note pads, and food gifts. Anything beyond that, however, becomes unnerving. You want to be sure there's no confusion of trying to buy referrals and that you adhere to local and federal laws in this arena. Since we're speaking of smaller gift items, if you have the means, try this out. You can see how effective it is and decide if it's the right approach for you. However, don't let the candy confuse the purpose of your visit. Just because you give out a small gift doesn't mean the nurse or receptionist has to get you access to the Doctor. Remember, you are differentiating yourself from the myriad of salespeople who visit physicians regularly. Stay dignified.

This seems like a good time to tell you my favorite question to ask Doctors when we're talking shop. If given the opportunity, I like to ask, "What are your Physical Therapy needs?" It's interesting to hear these responses. One doctor responded that he didn't want patient notes to be more than 1 type-written page. Another said that he didn't want us talking about leg length discrepancies with his post Total Hip Arthroplasty patients. Still another said that he wanted more notes from the office. All of this feedback was so helpful and it came from such an easy question. Remembering each Doctor's requests can go a long way in gaining them as a referral source.

Other good questions for doctors:
- How often do you refer to Physical Therapy?
- There are a lot of great Physical Therapy offices in this community. Are you happy with the medical care in this community?
- Are there any diagnoses that are challenging to find a good PT?

While you're dropping off Thank You notes, you can also give out flyers for your upcoming class.

Day 11 Checklist
- ☐ Go through Active Patient files and update status.
- ☐ Execute your weekly marketing plan.

Day 12

Today we're going to initiate a system of sending patient feedback to physicians. Thus far, we've been the ones sending notes to doctors, thanking them for their support and confidence. "Thank you" goes a long way. Now we're going to initiate a way of getting your patients to thank their doctors for sending them to your clinic. Imagine how powerful it is to have your patients return to their doctors and say, "Thank you for sending me to Best Physical Therapy. I feel so much better." Your patients probably tell you that they are so grateful for your help. Let's get that feedback to their doctors.

Appendix 11 shows a sample postcard that you can ask your patients to fill out. Tell patients you'll be happy to mail it for them. Your local printer can create postcards for you. TCG can also customize postcards for you. Contact us for assistance. At the midpoint of Physical Therapy, you should also ask your patients to fill out a survey, shown in Appendix 12. Send this survey, along with a letter to the patient's doctor. The letter should say, "Dear Dr. Jones. Here is what your patient is saying about Physical Therapy." The amazing feedback and testimonials you will receive is priceless. They are phenomenal marketing tools.

Day 12 Review:
- ☐ Send Thank You postcards and Surveys to Doctors.

Day 13

It's not uncommon to lose patients at the 2-3 week mark. Either they start to feel a little better and decide not to continue with Physical Therapy, or they don't feel any better and don't want to waste their time. To keep both ends of the spectrum coming to Physical Therapy, we're going to implement phone calls to patients at the 2-3 week mark by the Patient Liaison. Remember the PL is your first contact on the phones. The PL most likely scheduled your patient's Initial Evaluation. Now they're going to call at the 2-3 week point to see how the patient is progressing. Please see Appendix 13 for a script of what the PL should say on this phone call.

I began doing these phone calls with a well established office as a test. Truthfully, I thought these phone calls might be a waste of time. I couldn't have been more wrong. Patients were giving the PL such amazing information that I regretted not implementing it sooner and for all the offices I've worked with. We received so many compliments which are so nice to hear and reinforce our good habits. Of course, every now and then we heard something we didn't like. For example, we heard about staff members who rushed treatments, and scolded patients for being late for appointments. More commonly, however, the rare negative feedback was from patients who felt they weren't progressing and were bored with Physical Therapy. As soon as we heard this we addressed it immediately during the next PT visit. The PL is a little removed from the situation and for that reason, patients feel comfortable confiding in her. Trust me, this won't waste your time. If you hear 99 compliments and 1 complaint, it's still worth it so you can address any problems.

Some software programs can run a report for you that tells you when patients are at their two week mark. Otherwise you might have to implement a manual method of sorting.

Since we're going to be making phone calls regularly to patients, now is a good time to remind you to brush up on your HIPAA guidelines so you and your staff best protect patient privacy. Contact TCG with any questions about this.

Day 13 Checklist:
- ☐ Train PL to call patients 2-3 weeks after their Initial Evaluation to check their progress.
- ☐ Review HIPAA guidelines with staff to best protect patient privacy when making phone calls.

WHAT DO PATIENTS CONSIDER WHEN GOING TO PHYSICAL THERAPY?

Patient Decision Tree

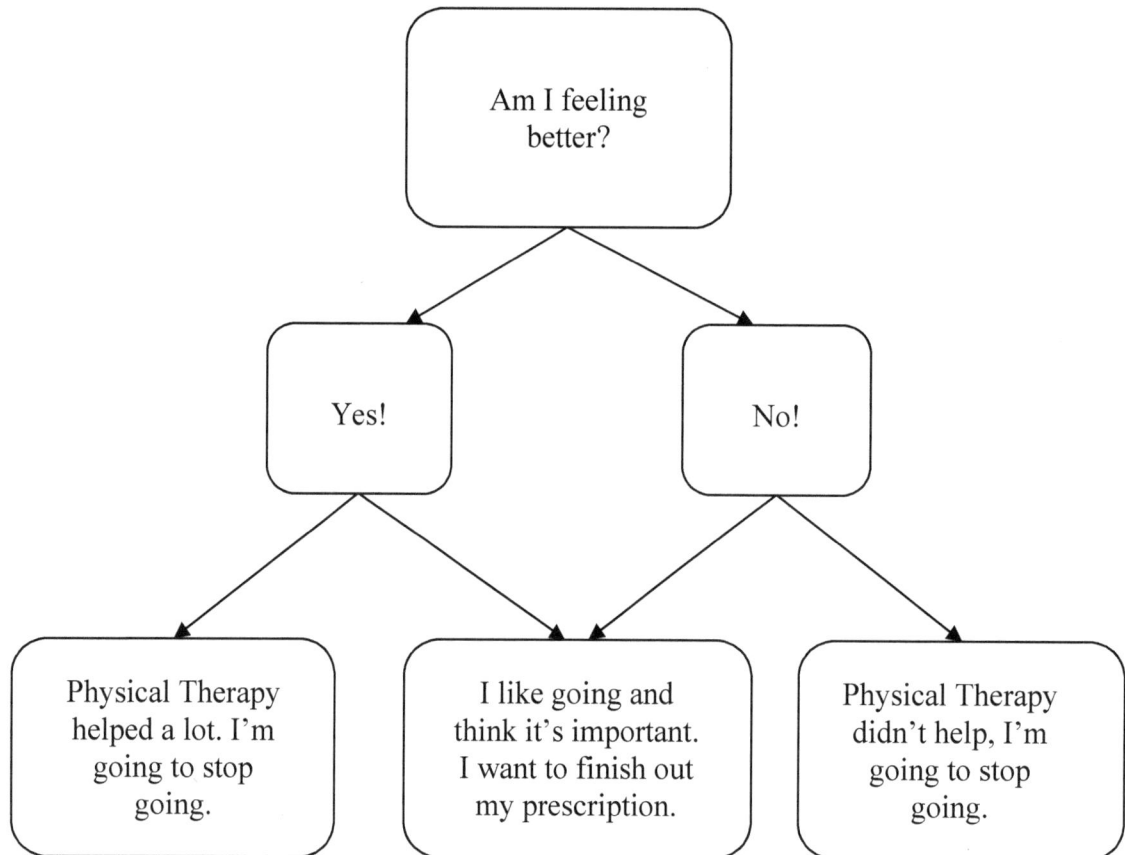

```
                    ┌──────────────┐
                    │  Am I feeling │
                    │    better?    │
                    └──────────────┘
                     /            \
              ┌────────┐        ┌────────┐
              │  Yes!  │        │  No!   │
              └────────┘        └────────┘
               /      \          /      \
    ┌──────────────┐ ┌──────────────┐ ┌──────────────┐
    │ Physical      │ │ I like going  │ │ Physical      │
    │ Therapy       │ │ and think it's│ │ Therapy       │
    │ helped a lot. │ │ important.    │ │ didn't help,  │
    │ I'm going to  │ │ I want to     │ │ I'm going to  │
    │ stop going.   │ │ finish out my │ │ stop going.   │
    │               │ │ prescription. │ │               │
    └──────────────┘ └──────────────┘ └──────────────┘
```

The diagram above illustrates the importance of keeping in touch with your patients. As you can see, if a patient is feeling better but the experience of Physical Therapy is a positive one, they are more likely to finish out their prescription than if they view Physical Therapy as a hassle. Similarly, if they are not seeing progress, they may discontinue coming to Physical Therapy. For this reason, it's important to keep communicating with your patients and discussing expectations.

Day 14

There's a lot of talk these days about treating the whole patient. 4-6 weeks of Physical Therapy is wonderful, but then what? Hopefully your patients feel so great that they want to keep up their workouts at home. Sometimes a good Home Exercise Program (HEP) is enough. If not, execute a Step Down program for your clinic. Do you have the space for patients to come to your clinic and workout for a monthly fee? This is a great way of getting some extra cash pumping into your office. You can have your patients sign a waiver and workout on less busy days, for example, Tuesdays and Thursdays between 1 and 4pm.

Another great idea is running a class. A lot of PT clinics have jumped on the Pilates bandwagon and teach regular classes. Others do a Fit after 50 class. You can teach whatever you are comfortable with. Consider making an exercise class as an offshoot of your public speaking. For example, if your free class to the community is on avoiding knee injuries, teach an exercise class of stretching and plyometric drills geared towards this population. Any class is a value added service that make your patient's experience at your clinic more complete.

You can also consider renting space to a Personal Trainer. Typically the Personal Trainer will give you a percentage of what he brings in, or a flat rental fee. Your patients may really like this option.

Unfortunately, most clinics don't have the space for a step down program of any kind. If you don't, contact the nearest gym with the best reputation. Discuss setting up a Step Down program with them. Would they be interested in offering a discount to your discharged patients? In return for your referrals to them, would they recommend your clinic for their clients' Physical Therapy needs?

Day 14 Review
- ☐ Establish a Step Down program in your clinic or at a local health club.

Day 15

Hooray! You made it to the midpoint of the program. If you started this program on a Monday, then today is Friday and you have two regular tasks to do. First, look back on the schedule and make sure that everyone is scheduled for next week. Next, create your Trend Report based on your Marketing Surveys and establish next week's marketing plan. When you are finished, your final task of today is to write a newsletter for your practice. This task is purposely assigned for a Friday so you have the weekend to complete it. Nearly every word processing program has a template for newsletters. Take your time and find the layout that works best for you. It doesn't need to be longer than 1 page, unless you prefer something longer. TCG can also help you out with this task. We won't worry about distributing the newsletter until Monday.

Almost every practice these days has a newsletter and they're all starting to look the same. The same boring article generator is spewing out these newsletters for practices nationwide. These newsletters are boring to say the least! And I doubt many people are reading them. Here's what I recommend: make your newsletter personal. Your patients know you and your staff. They care about you, like you care for them. Give them updates like who got married, who had a baby, and who earned a degree. When you have a staff function, like a holiday party or birthday, take pictures and put those in your newsletter (as long as they're appropriate!). Treat these newsletters as though you are writing to friends and your patients will appreciate it. You should also provide information on any upcoming classes and services you provide. Put your name and tag line in a very obvious location.

I have to share a story. I worked with a practice for many years and wrote their newsletters. I wasn't sure anyone read them until one month when I left out the recipe that I usually included. Boy, did I get complaints! The recipes were a big hit because they were personal recipes, typically favorites from the staff. Patients loved them. Incidentally, 20% of new patients in that clinic were Reactivations. Obviously they were doing a good job and staying in the forefront of their patient's minds. I think their monthly newsletters played a big role in that.

In today's day, you can email newsletters to your client list. If the majority of your patients, however, do not provide an email address, you should consider traditional mail for the newsletter. You want your newsletter to reach as many former patients as possible. The frequency of the newsletter is up to you. I recommend monthly or quarterly.

Day 15 Checklist
- ☐ Go through last week's schedule and make sure that everyone is scheduled for the upcoming week. Call those who are not scheduled.
- ☐ Create your Trend Report and Marketing Plan
- ☐ Create a newsletter for your practice

Week 3 in Review

You're at the halfway mark. How does it feel? At this point you've invested a lot of time and energy into your clinic. It might be very tempting to give up, but try to stay the course! This program is designed to be completed in 30 days. It's intense but gets results quickly and affordably. This past week you did the following:

- You delivered notes and visited referring physicians.
- You trained your PL to call patients after 2-3 weeks of treatment.
- You developed a step down program in your clinic or with a local gym.
- You wrote a newsletter for your office.

In addition your office is now:

- Distributing New Patient Packets
- Going through patient files each Monday
- Checking last week's schedule against the upcoming week's schedule every Friday
- Creating a Trend Report and analyzing statistics
- Calling patients after their Initial Evaluation
- Calling Case Managers after their patient's Initial Evaluation
- Sending out Postcards and Thank You surveys to doctors.

There are 3 more weeks to go. Keep it up!

Day 16

Week 4, here we go!

After you have reviewed your Trend Report, you should have your marketing assignment. Write up any cards, prepare your notes and get them out to the doctors on your list. Remember, you can always refer back to Appendix 5 for samples of Thank You Notes.

Today you should also get your newsletter out. Decide if you want to email it or send it in traditional mail. The advantage of email is, of course, cost. The advantage of traditional mail is that you may reach more of your client base depending upon the demographics of your patients.

Finally, since today is Monday, be sure that someone checks all the active patient files and updates their status. Is anyone discharged? Does anyone need a phone call?

Day 16 Checklist
- ☐ Execute marketing plan.
- ☐ Send out newsletter.
- ☐ Go through active files.

Day 17

We're going to involve your staff for today's assignment and go through any existing advertising materials that you have for your clinic. This means, you should collect your letterhead, business cards, brochures, folders, exercise sheets, and flyers. Ask your staff to give you feedback on your existing materials. Does anything need a facelift? Are there typos? You want to be sure that anything going to referral sources looks the best it can. Ask for feedback on the New Patient Packets. Are patients commenting about the packets? Are there additional materials you want patients to have about your services? Decide what you want to create or change. Please distribute Appendix 14 to your staff and get some feedback on your marketing materials.

Next let's focus on your website. In today's world, you have to have a website to be competitive. We go online for everything. Your website needs to present you in the best possible light. If you don't have a website, today's task is to create one. Some browsers allow you to make a website for free. We're going to focus on basic concepts in web design. Once again, please use Appendix 14 and gather feedback from your staff.

Please note that website design and marketing is a very big field and is very specialized. We could spend days talking about it! If you want more information about professional website design and marketing, contact TCG for details and referrals.

Think of today as a day devoted to refreshing your image. You get your hair cut regularly to maintain a certain image. Similarly, you need to refresh your marketing materials to maintain the image of your business. Don't assume that you need to spend a lot of money in this area. You just need to keep things current, professional, and informative. For example, I know a phenomenal Physical Therapist whose website, in June, is still talking about snow cancellations from January on his home page. If I'm a consumer looking at that home page, I'm not getting the most current information I need. Consumers only spend seconds on a website and they don't want to hunt for information. You can't afford to be lazy with your website. Keep your website up to date and your materials current so your practice always has its best foot forward. Remember to think of your time as money and lost opportunities as lost income.

It is highly recommended that you also create a Facebook page for your clinic. Facebook is one of the biggest marketing tools out there. It is very user friendly, but if you need help creating a page, TCG (or most teenagers) can help you. Suggest your business page to patients, friends, and family. If you have a personal page and you're on Facebook all the time, be careful, and keep your profile private. You should be on Facebook as a company, not clinician. You may not want those with whom you have professional relationships with seeing pictures of you tagged from last month's bachelor party or reunion. Remember, you are crafting your image and building your reputation. Keep as much as possible in your control. Use your Facebook page to advertise classes, post greetings, and keep in touch.

Day 17 Checklist
- ☐ Review written marketing materials for typos, content, and appearance
- ☐ Create or review website.
- ☐ Create or review Facebook page.
- ☐ Make changes to your materials as needed

Day 18

Remember our three circles of referrals – External Referrals, Internal Referrals, and Reactivations. Internal Referrals can be a huge source of income for you. This means a patient is coming to your clinic because their friend or family referred them. When you find a great product, don't you like to tell others about it? Healthcare is no different. Your product is Physical Therapy. Rehabilitation is such a personal experience. People are definitely talking about their injuries to their friends and family. Hopefully they're saying good things about your practice, despite their pain and suffering. Today we're going to put a Referral Rewards Program in place to thank your patients for referring their friends and family to you.

The Referral Rewards Program is very specific. If any patient, past or present, sends you another patient, send them a Thank You note. If they send another patient after that, send a note with a little gift. I recommend that this gift be a gift card to a local establishment, for example, a coffee shop or restaurant. If they continue sending patients, you can increase the amount of the gift card or keep sending the same ones. Try to avoid franchises or chains. The reason why I'm stressing to keep things local is to establish relationships with other local businesses. By buying gift cards you're sending them business and hopefully they will return the favor and refer to you whenever they can. If you buy gift cards in bulk, they may be able to give you a discount. Think carefully before choosing a restaurant. Is there somewhere you go for lunch regularly or get your coffee each day? Do you personally know the owners of any local business? Have you treated any local business owners or employees?

The amount of the gift is less important than the act of acknowledging the referral. You will find out from your Marketing Surveys the names of former patients who are referring. They probably won't be expecting a thank you note or a gift from you which makes what you're doing that much nicer. I worked with a practice that boasted 50% of all new patients from Internal Referrals or Reactivations. This is tremendous! The customer service at this practice was so exceptional, that people couldn't help but talk about it and they hated to leave. If 50% of your new patients are repeat customers or internal referrals, then you don't have to invest much in external marketing efforts. You also have a nice cushion if one of your physician referral sources retires or starts his own PT practice.

Today's second task is for your PL to make patient calls to patients who are 2- 3 weeks into treatment, as we discussed last week. This task will be done every week.

Day 18 Checklist
- ☐ Setup Referral Rewards Program by buying gift cards from a local business.
- ☐ Have PL call patients who are 2-3 weeks into treatment.

Day 19

We're going to talk about the holidays today. Remember, we always want to be in front of referral sources without being obnoxious. Visibility and repetition is so important. How many times do you need to see a commercial before you remember it? How many flyers and junk mail do you throw out without reading, and what actually catches your eye? What you notice depends on your mood and your needs at the time. You have to stay in front of referral sources with consistent advertising. Referral sources can be physicians, community groups, businesses, and of course, former and current patients. So far, we've talked about:

- Writing thank you notes to Doctors
- Hand delivering patient notes
- Hand delivering treats or branded items to Doctor's offices
- Scheduling free classes for the community in your clinic and in town
- Keeping up with patients via phone calls
- Sending out newsletters to all current and previous patients
- Asking patients to send postcards or surveys to their doctors
- Updating and maintaining your website and Facebook profile.

Today we're going to look at the holidays and present options for how you can use holidays in your marketing campaigns. Depending upon when you're doing this program, a holiday is surely approaching. Consider how you will use the holiday to stay in front of Doctors and Patients.

Holidays in Chronological Order

January: There are a cluster of winter holidays – Hannukuh, Christmas, and New Year's Day - that are often combined. Use New Year's Day to send out an email wishing everyone a happy, healthy, and prosperous New Year. If you have a Pilates or some other exercise class, use this time of year to increase advertising and capitalize on healthy New Year's resolutions. Print up flyers and hand them out. Send out emails, post on Facebook, and talk up your programs for getting fit and feeling your best. Your newsletter should talk about being cautious around snow and ice, fall prevention/ balance training, and help for back pain. Winter is a great time of year to schedule a free class in your community. Shoveling and lack of activity this time of year often exacerbates injuries.

February: Love is in the air! Your message to your patients is very simple this month: We love taking care of you! Thank you for letting us do what we love. Happy Valentine's Day to you and your loved ones. To Doctors: Thank you for your referrals and for giving us the opportunity to do what we love. Happy Valentine's Day to you and your loved ones! Send this message via email or cards.

March: Wish everyone the luck of the Irish in March. Consider passing out Irish Soda bread or green bagels to your best referral sources or to improve relationships with certain referral sources. Also, get some green goodies to share in your clinic with your staff and current patients. Everyone is Irish on St. Patrick's Day!

April: Nurse's Week is in April. You would be an April fool to miss this holiday! Pass out a card – preferably in person – to every Nurse you see. Nurses are very commonly the Referral Managers in most offices. Don't ignore them, especially on their holiday.

May: May brings Mother's Day and Memorial Day. Use your newsletter to wish Mom's a Happy Mother's Day and to thank Veterans for their service. Send out an email to everyone wishing them a happy, healthy, safe, and enjoyable Memorial Day weekend. Post thanks and well wishes on Facebook.

June: Don't forget Father's Day in June. Wish everyone a Happy Father's Day in your Newsletters and online.

July: Once again, send everyone an email wishing a happy and healthy holiday for the 4th of July. Since a lot of people are traveling over the summer months, you may want to send out a reminder to current patients about coming to Physical Therapy and letting you know when a vacation approaches so you can plan your treatment program. Remind them that there is no vacation from Physical Therapy and they should resume their appointments when they return from vacation.

September and October are great months to schedule a class. People will be back from their summer vacations and back in the swing of school, and work. You should expect good attendance. Remember to advertise for your class on your website.

November: Everyone sends Holiday greeting cards. It's a great idea to beat the rush by sending out Thanksgiving cards. This way your card is first and stands out. Your message is simple, "Thank you for your support and confidence in all of us at Best Physical Therapy. We wish you and your family a happy and healthy Thanksgiving."

December: You can choose to do a "Seasons Greetings" card or email in December. People get so many of these. I think a great way to stand out is take a picture of your staff and say, "From our family to yours, we wish you a Happy and Healthy Holiday Season." This card, or email, can go to referral sources and patients, past and present. The picture should also be posted on your website and Facebook pages.

Day 19 Checklist:
- ☐ Plan how you will use the next holiday to stay in front of your external and internal referral sources.

Day 20

You should always be able to offer some kind of freebie for prospective clients to get them in your doors. For example, if you have a cold laser program for cash, you may want to offer one treatment for free. One clinic I knew had a water massage bed and offered free 15 minutes massages. Another clinic offered a Free Consultation with a Physical Therapist. The consultation included a free book and a fixed amount of time with a Physical Therapist. Other ideas would be a 15 minute chair massage, free attendance to a Pilates class you're giving, or Free paraffin hand treatment. Today we're going to create a coupon for the freebie you're going to give away. Because it is the least expensive, we give an example of a Free Consultation coupon in Appendix 15. Please consider putting a Freebie coupon in your New Patient Packet to be used by someone other than the new patient.

Be sure that your entire staff knows how to schedule the Freebie you are advertising. For example, if someone calls and asks to schedule the "free massage session" you are advertising, be sure your staff knows to schedule a 15 minute massage, and not a whole session.

It's Friday again! Be sure to do your Trend Report. Also, please review the schedule and compare it against next week's schedule to see if any patients are missing. Finally, remember to review the Trend Report and create your Marketing Plan.

Day 20 Review
- Create Freebie Coupon
- Create Trend Report and Marketing Plan
- Review patients on last week's schedule against the upcoming week

Week 4 in Review

This week we worked on your online presence by auditing your website and Facebook page. We setup your Referral Rewards Program and your clinic Freebie. We also talked about maximizing the holidays to connect with patients.

By now you should be getting accustomed to the tasks that are asked of you each week, such as going through charts to check for patients who are not scheduled, and constantly checking the schedule. Your office should also be getting comfortable calling patients after their first visit, and reviewing the New Patient Packet. All these details remind patients how important Physical Therapy is to their health and well being.

Day 21

Week 5 is here! You're almost at the finish line. This is the last big stretch. Don't give up! Today we execute our marketing plan based on the Trend Report and consultation with TCG. We also give out Referral Rewards to patients who have recommended your office to their friends and families. Finally, have someone go through all the active patient charts and update patient status.

Day 21 Review
- ☐ Execute marketing plan
- ☐ Give out Referral Rewards based on Marketing Surveys
- ☐ Update patient status on all active charts.

Day 22

I've heard that retailers know a previous customer is more likely to buy from their store than a new customer would be. That's why stores spend so much time and effort mailing you circulars and emailing you once they have your information in their system. You shopped there once and you're likely to shop there again. Remember healthcare is something we shop for now. Our past patients are shoppers who might buy from us again if they have a need. To stay in their minds, we're going to setup a system of calling discharged patients to see how they're doing. They may have reinjured or not fully recovered and need to return to you. Or they might talk to you about a new injury or the new services you provide. You can enlist the PL to make these calls or ask the entire staff to make a certain number of these calls per week. Your phone call should express the following,

"Hi Mr. Smith, this is Terri from Best Physical Therapy. You came to see us back in November and I'm calling to see how you are feeling. I hope you're doing well. If you have any questions or Physical Therapy needs, please don't hesitate to contact us."

Some patients may use the opportunity to talk about a medical issue. Be sure the person who makes these calls is trained enough to know that they can't dispense medical advice. One office I worked with put an inexperienced PT Aide on the phone to make these calls. A patient who had exhausted PT benefits tried to say that the Aide guaranteed that he could come in for Physical Therapy despite having no insurance coverage. Another time, I heard an Aide telling a patient on the phone to add more weight in her Home Exercise Program. Even though this was the Aide's opinion, he is representing your office when he makes the call. He should know what he can and can't say. That's why even though these calls are pretty easy to make, a skilled worker is the best one to make them.

The practice I mentioned that had a 50% rate of Returning Patients and Friends and Family referrals used this technique and I think it really helped. A good way of getting your PL and Physical Therapists to do this task is to make it mandatory that they make a certain number of phone calls per week. For example, everyone on staff has to make 10 phone calls per week. Discharged patients should receive calls at the 30-60 day mark, the 6 month mark, and the one year mark. Calls can be made once a year after that.

Now that you are regularly updating the status of your patients, you should be able to come up with a system of recording the discharge date of your patients. A lot of software programs automatically run date of discharge for you. On days that you are slow, encourage your staff to make more calls. A bonus gift card for whoever makes the most calls each month can be a good motivator.

Day 22 Checklist
- ☐ Setup a system where you call patients 30-60 days after discharge, 6 months after discharge, and 1 year after discharge.

Day 23

Remember those Freebie coupons you worked so hard on? Today you're going to canvas your neighborhood and give them out wherever you're allowed. Think back to any old friends or former patients. Can you leave Freebie coupons in their places of business? You should also go on your town's website and contact the following:

- Religious organizations
- Women's groups
- Senior groups
- Active Adult Communities
- Adult Sports Leagues
- Local Coaches
- Local Athletic Trainers

Religious groups and other community groups may be interested in your Freebie to help people in need, to educate others, or to help as a donation. While you have them on the phone, you should also ask if they would like you to present a free Lunch-and-Learn presentation. Public speaking can be challenging, but the more talks you give, the more you will be thought of as an expert in your field.

If you work in a town that you grew up in, your high school may allow you to give a presentation of some kind to students, parents, and/or faculty. If you are allowed, give out Freebie coupons. A Physical Therapist I know built his practice this way. He went back to his high school and talked to everyone he could. He quickly became the recommended Physical Therapist for all the student athletes. After a while, it wasn't just the athletes coming in for treatment, but their parents and friends and families. He has two locations now and his staff is growing. The lesson here is that a small lead can turn into something great if you nurture it.

Today's other task is for your PL to make calls to patients at the 2-3 week mark of their Physical Therapy program.

Day 23 Checklist
- ☐ Distribute Freebie Coupons
- ☐ Have PL call patients 2-3 weeks after their Initial Evaluations.

Day 24

Today you're going to continue hitting the pavement with your Freebie coupons. Your first stop is your nearest hair salon. Yes, you read that right. You're going to give your coupons to the salon and be as friendly as possible to every hair stylist you see. Here's why, hairdressers stand and do repetitive tasks all day. They're hard on their bodies. Their back, feet, and hands are often achy. The second reason is because hairdressers know everyone in town. They meet and talk to local residents all day long. Also, most of the clients at a salon will be women and women typically make the health decisions for their families. If a hairdresser likes your services, you can't ask for a better billboard. One of the towns I worked in had 3 hair salons on one street and they all did a fair amount of business. Don't limit yourself to one salon. Stop by all of them and network, network, network!

Think back to your patients. Have you treated a local policeman, fireman, or local business owner? Call them and ask if you can stop by with some Freebie coupons for them and their coworkers. You may want to give them a thank you note and Referral reward for their trouble. Think back to Day 8. Did you make any contacts in the Town Hall? Can you go back there with your coupons? Distribute your coupons wherever you can but be careful about hanging them in public areas. Every town has different laws about that.

Day 24 Checklist
  ☐ Distribute Free Consultation Coupons at Hair Salons and local groups and businesses.

Day 25

There are a number of things that I ask you to do each week to have a successful, efficient practice:

- Update the status of all patients
- Call patients a day or two before their Initial Evaluations to remind them of their appointment date and time
- Call Patients after their Initial Evaluation
- Call Case Managers after Initial Evaluations
- Call patients 2-3 weeks into treatment
- Give out Referral Rewards
- Create Trend Report based on marketing surveys and
- Call discharged patients
- Send postcards and surveys to doctors
- Develop and execute a marketing plan based on the Trend Report

And each month:

- Send out birthday cards
- Write and distribute a newsletter

Today you should spend a few minutes with your staff and work on creating Job Descriptions. Distribute Appendix 16 to each employee and have them list all the tasks they are responsible for. After each staff member completes their sheet, review it with them privately. Consider the following:

- Does each task have a frequency, such as once a week, twice a week, etc?
- Are there tasks you want to reassign to someone else?
- Are you satisfied with what is getting done?
- Do you want to add more responsibilities?
- Are these tasks an efficient use of time?
- Is the employee comfortable with this job description?
- Is the employee capable of doing all that has been assigned?

This assignment has many purposes:

1. You will learn exactly what your employees are doing.
2. You learn what your employees see as their main responsibilities.
3. You get to reassign job tasks.
4. You establish accountability for jobs.
5. You learn what you expect from each position.

Similarly, your staff learns what is expected of them.

After you have reviewed these sheets, you should both sign the bottom of the page. You should go back to this document from time to time and update it. I recommended reviewing it at least every quarter. If you hire any new employees, write a job description/task sheet up for them and give it to them on their first day. As their position evolves, make the necessary changes to this document.

Today you should also do your Trend Report based on your Marketing Surveys and schedule a meeting with TCG to discuss findings. Finally, compare last week's schedule against the upcoming week's schedule to be sure no one slipped through the cracks. Call anyone who isn't scheduled and should be.

Day 25 Checklist:
- ☐ Create Job Descriptions with every member of your staff.
- ☐ Create Trend Report based on Marketing Surveys and develop a Marketing Plan
- ☐ Compare last week's schedule against the upcoming week.

Week 5 Review

5 more days to go! You should be proud of all that you have already accomplished. If this intense focus on your business is uncovering new challenges that you want to work on, don't despair. Keep a log of new projects and set realistic time frames for them. You'll have the practice of your dreams in no time.

This past week we initiated some new programs such as:
- Referral Rewards
- Discharge Calls
- Freebies

Referral Rewards and Discharge Calls help solidify your reputation as a caring practice. Both are designed to go the extra mile and put the personal care back into healthcare. Freebies are a nice way to get people to come into your office who might not be aware of your services.

Day 26

Today is your last Monday! Today you should meet with any employees that you didn't get to on Friday and go through their Job Description/task list. The rest of today's tasks have all been done before:

- Give out Referral Rewards
- Go out on Marketing stops based on Trend Report and Marketing Plan
- Go through charts and update status of active patients. Call anyone who hasn't been to Physical Therapy and isn't on hold.

Day 26 Checklist:
- ☐ Referral Rewards
- ☐ Execute Marketing plan
- ☐ Update status of each active chart and make calls.

Day 27

Cancellations and No-Show's are costly for your practice. When you have a Cancellation or a No-Show, you miss out on the opportunity to schedule another patient. This opportunity cost can be expensive for your practice. You need to encourage attendance at your clinic in an appropriate way. You may already charge a cancellation fee. The decision to use a cancellation fee is up to you. It can be an effective way to discourage cancellations and make up for a cancellation on the back end. Or, you may see it as an uncomfortable burden on your patients.

Your PL should be going over your Cancellation Policy on the patient's first day. Instead of only focusing on cancellations, we're also going to try to encourage attendance by establishing Perfect Attendance Rewards for patients. If a patient comes in for 4 weeks without any cancellations or No-Show's, give them an item with your logo – their Perfect Attendance Reward. This could be a mug, T-Shirt, sticky pad, or magnet. Tell patients about the contest on their first day and show them the reward they will receive. Also explain the difference between a cancellation and a re-schedule. The difference is that a re-scheduled appointment is made up within the same week. For example, if I have appointments for Monday, Wednesday, and Friday, and move Monday's appointment to Tuesday, that's a Reschedule. If I cancel Monday's appointment and don't make it up, that's a Cancellation. Your staff should already know the importance of trying to reschedule every cancellation. They should be offering later appointment times and trying to recover the visit. Use the Perfect Attendance Award to reinforce the importance of coming to Physical Therapy.

Your staff should also get into the habit of telling a patient when their next appointment is whenever they leave the office. "Goodbye Mr. Smith. We'll see you on Wednesday at 1:30." This is a great reminder for the patient.

Day 27 Checklist
- ☐ Initiate Perfect Attendance Rewards

Day 28

Today we're going to prepare birthday cards for your patients celebrating birthdays next month. Once again, this is a way of staying in front of patients and reminding them that we care. You can buy birthday cards that are personalized for your office, or buy generic ones. You can also send out emails. Most likely, you will have to prepare a combination of cards and emails based on your patient population. Your cards and emails should be as personal as possible. Try to have a signature on the cards and include a personal line or two in the emails. Obviously this is a task you'll delegate to someone else on staff. Have them show you who is celebrating birthdays each month so you can dictate a personal greeting, if appropriate.

A lot of computer programs will generate a birthday list for you. If that's not possible, leaf through charts and start keeping a birthday list.

Day 28 Checklist:
☐ Prepare and send out birthday cards or emails

Day 29

Focus today on getting a firm grasp of the Discharge Call list. Have your PL make calls with you. Anyone making these calls should be extremely friendly and polite. The challenge of these calls is to make them brief and not get caught up in a topic. Practice doing a few yourself so you are familiar with the procedure that you want executed.

Today is a great day to also look back over the tasks you've done since beginning this program. Do you have any questions or concerns about anything? Contact TCG for guidance.

Day 29 Checklist
- ☐ Make calls to discharged patients.
- ☐ Review what you have done since beginning this program and contact TCG for the answers to any questions.

Day 30

You have finally made it to your last day. Congratulations! Today is a fun day. Today you are going to hold "Patient Appreciation Day!" This day is about thanking patients for letting you do what you love and celebrating your clinic with them.

You may consider trying some of these fun things today:

- Leave a small snack like vegetables, a fruit platter, or popcorn out for patients.
- Ask patients to leave their business card for a chance to win a prize. (Use a gift card from your Referral Rewards stash as the prize.)
- Give out items branded with your logo like pens, hats, and t-shirts to everyone who attends Physical Therapy today.
- Play movies on a TV while people are going through their exercises.
- Make coffee in the morning for your morning patients.
- Give out Freebie coupons to everyone to share with their friends and family.

Patient Appreciation Day can be held as often as you like. Some clinics hold it once a month, others once a quarter. Every other month is fine, too. This day really helps patients walk away with a good feeling about your clinic. It also lightens the mood in your clinic which can be important some days. I recommend holding it every month with few exceptions.

Finally, since today is Friday, get your Trend Report done and develop next week's Marketing Plan. Also, check last week's schedule against this week's. It's the last time, I'll be reminding you!

Day 30 Checklist
- ☐ Hold Patient Appreciation Day
- ☐ Check last week's schedule against this week's schedule.
- ☐ Create your Trend Report and Marketing Plan.

Week 6 in Review:

This week you repeated your weekly tasks. You also initiated Perfect Attendance Rewards, Patient Appreciation Day, and Birthday cards. All three of these give special attention to patients. Unfortunately, the current healthcare climate has caused a lot of healthcare practices to lose their personal touch. You will stand out by being a practice that truly cares about your patients.

By now you might be seeing which tasks you can delegate and which you want to scrutinize. Next we will discuss ongoing maintenance of the 30 day program so that your clinic always shines.

Although you just finished the program, you may already be seeing a boost in business, or you may have noticed that the upcoming weeks are busier than ever. The effects of this 30 Day Program are not just seen within these 30 days, they are ongoing. Depending on the time of year, some clinics respond to this program very quickly, and others take a little longer.

You may also have noticed changes in your work environment. Your office should be more organized and your staff more devoted to patients. Please remind your staff to talk to you. Keeping the lines of communication open during times of growth and change is so important. You may also need to use bonuses to help stir motivation. For example, bonuses may go to:

- The clinician with the lowest cancellation rate.
- The receptionist with the highest reschedule rate.
- The staff member who makes the most discharged calls.
- The staff member who gives out the most postcards or discharge surveys.

Your last official task is not to let the ball drop. You've developed so many great habits during the course of these 30 days. Keep it up! Keep communicating with your office and doing your weekly tasks. Keep distributing your Marketing Surveys and analyzing your Trend Reports. Most importantly, keep having staff meetings and know what's happening in your office. Your focus and attention will pay off and you will have the practice of your dreams!

Recommended Ongoing Tasks

We spend a lot of time repeating certain tasks. The following jobs are recommended to be ongoing responsibilities that are completed each week:

Weekly:
- Update the status of every active patient making sure patients who are not scheduled are called or discharged.
- Call Evaluations a day or two before they are scheduled to come in to remind them of their appointment.
- Distribute New Patient Packets, explain policies.
- Create Trend Report.
- Visit 5-10 doctors a week based on the Trend Report. Hand deliver notes.
- Send Thank You notes to new Referral sources
- Call patients 2-3 weeks after their Initial Evaluation.
- Give out rewards according to Referral Rewards Program
- Call at least 5-10 discharged patients
- Give out Perfect Attendance Rewards
- Send out Thank you postcards or surveys to doctors

Monthly:
- Create a Newsletter
- Update website and Facebook page as needed.
- Send out birthday cards
- Decide if you want to do anything for the upcoming holiday
- Hold Patient Appreciation Day

APPENDIX 1

Staff Meeting Agenda

I.   We are going to begin a new program: 30 Days to More Patients and a More Organized Office
    a. Purpose: To increase business and run an efficient office
    b. Strategy: Different assignments given over 30 days that require daily completion.
    c. Based on several principles:
        i. There are plenty of referral sources besides Physicians
        ii. There are other marketing strategies besides buying lunch for doctors.
        iii. There are three main categories of referrals: Internal Referral Sources, External Referral Sources, and Reactivations, see Graphic 1.
        iv. Personal care and attention is so important, especially in this healthcare climate.
        v. Every patient has the potential to bring in new business, see Graphic 2
        vi. There are tasks that need to be done on the same day, every day, to keep the office running smoothly.

II.   Teamwork
    a. The success of this program relies on the participation of the entire team.
    b. Keep the lines of communication open
        i. We may experience growing pains as we get busier.
        ii. Talk about any problems as soon as they arise
        iii. Responsibilities may shift as we take on new tasks, and discard inefficient methods of doing business.
        iv. This is an exciting time to be a part of this company as we grow and improve our practice into the successful healthcare business we know it can be!

III.   Feedback
    a. Our first task is to review the aesthetics of the office
    b. Please fill out sheet (Appendix 2) as honestly as possible.

GRAPHIC 1

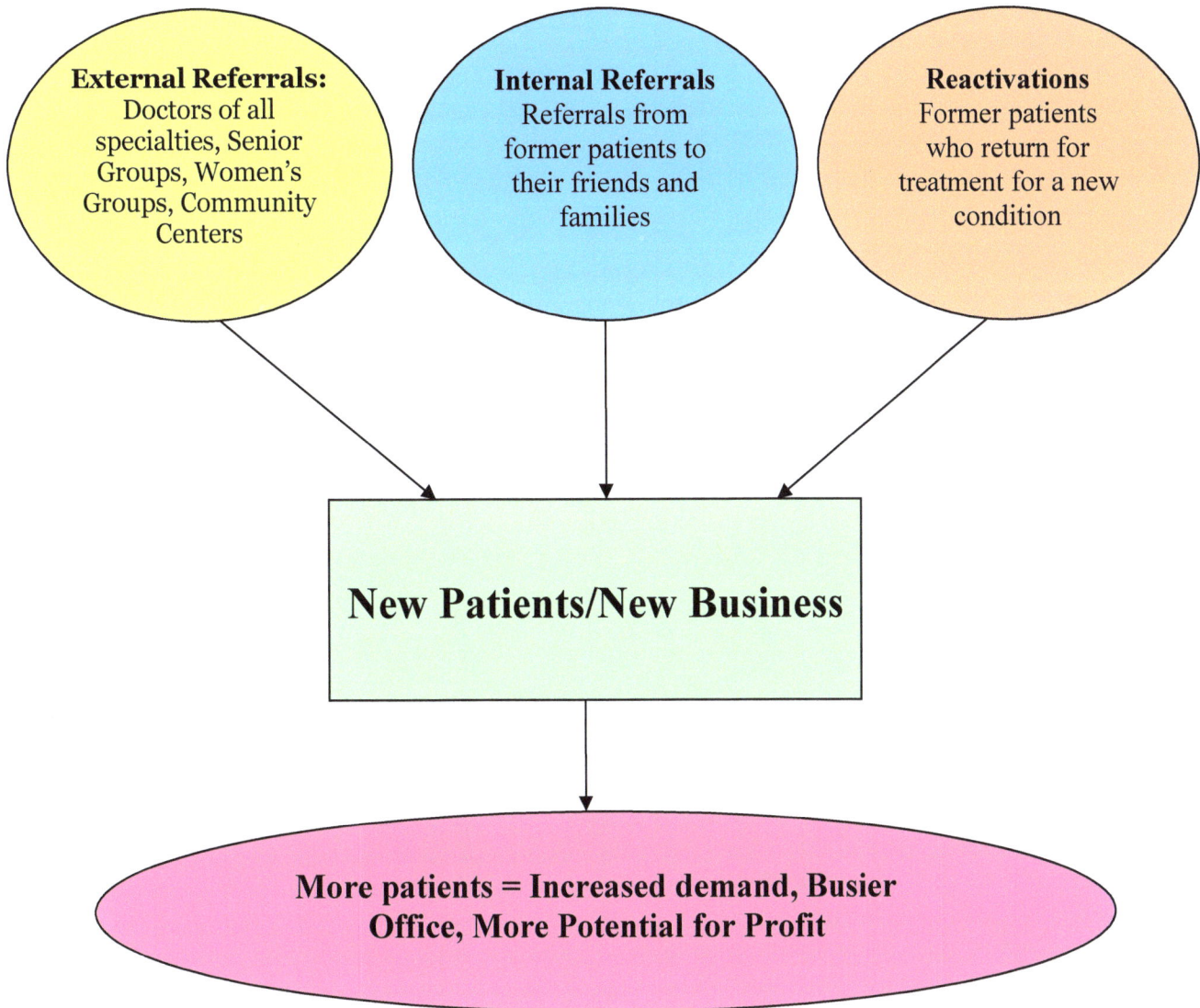

GRAPHIC 2

This graphic provides a visual explanation for how each current patient can bring new business to your clinic.

Question: Mr. Smith is your patient. How can he bring you more business?

His colleague is about to have the same surgery and Mr. Smith recommends you for PT.

He gets better right away and is discharged. His Doctor is impressed by his recovery and sends more post surgical patients to you.

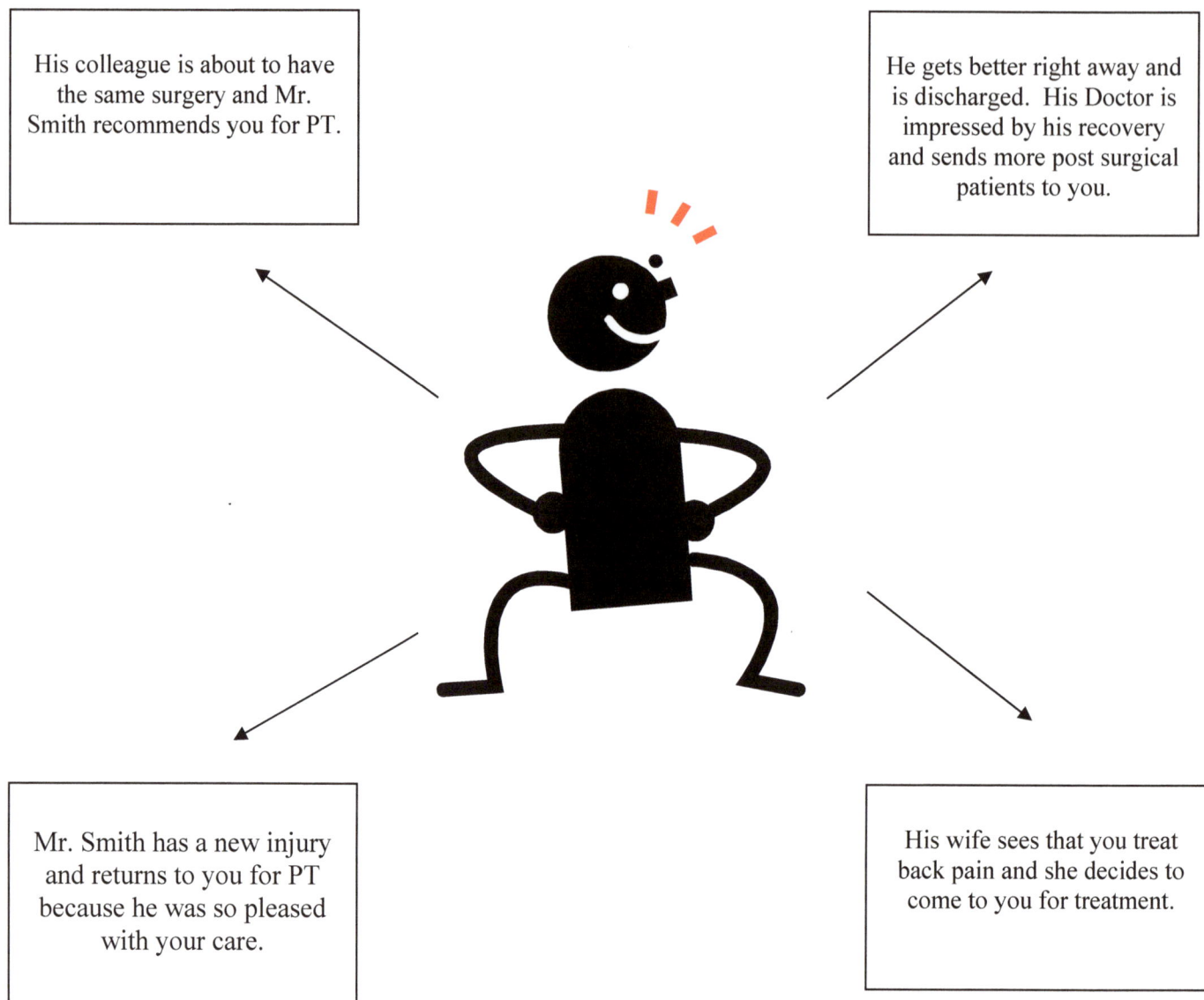

Mr. Smith has a new injury and returns to you for PT because he was so pleased with your care.

His wife sees that you treat back pain and she decides to come to you for treatment.

APPENDIX 2

Directions: Please exit and re-enter your facility. Fill out the following survey based on your observations and share it with your co-workers.

Please rate the following categories on a scale of 0-5 based on the following scale:
1 = Poor   4 = Good
2 = Unsatisfactory   5 = Great
3 = Satisfactory

	5	4	3	2	1	Comments
Visual appeal of signs outside the office. ▪ *Easy to see and read?*						
Cleanliness of check-in area.						
Cleanliness of waiting room.						
Cleanliness of Restroom.						
Efficiency of check-in process						
Appearance of exercise equipment. ▪ *Does the equipment seem outdated or old?*						
Appearance of mats and tables. ▪ *Are there tears on the mats or are they in good condition and clean?*						
Layout of gym and exercise equipment ▪ *Can you get to each piece of equipment easily?*						
Décor in the waiting area. ▪ *Is it inviting, soothing, and appropriate?*						
Décor in the office area. ▪ *Is there a lot of clutter that's distracting or is it organized?*						
Décor of the treatment area and gym. ▪ *Is it appropriate for a Physical Therapy office?*						
Appearance of staff. ▪ *Do you enforce a uniform or dress code?*						

APPENDIX 3

Directions: Please exit and re-enter your facility. Fill out the following survey based on your observations and share it with your co-workers.

Please rate the following categories on a scale of 0-5 based on the following scale:

1 = Poor   *4 = Good*
2 = Unsatisfactory   *5 = Great*
3 = Satisfactory

	5	4	3	2	1	Comments
Visual appeal of signs outside the office. ▪ *Easy to see and read?*						
Cleanliness of check-in area.						
Cleanliness of waiting room.						
Cleanliness of Restroom.						
Efficiency of check-in process						
Appearance of exercise equipment. ▪ *Does the equipment seem outdated or old?*						
Appearance of mats and tables. ▪ *Are there tears on the mats or are they in good condition and clean?*						
Layout of gym and exercise equipment ▪ *Can you get to each piece of equipment easily?*						
Décor in the waiting area. ▪ *Is it inviting, soothing, and appropriate?*						
Décor in the office area. ▪ *Is there a lot of clutter that's distracting or is it organized?*						
Décor of the treatment area and gym. ▪ *Is it appropriate for a Physical Therapy office?*						
Staff recognition. ▪ *Is there a way to distinguish who works at this facility – by name tags,*						

uniforms, etc?
Staff appearance
   ■ *Does the staff appear professional?*
Welcome
   ■ *Are you immediately and warmly greeted upon entry?*
Phone greeting
   ■ *Are you greeted in friendly tones and helped right away?*						

How many times did the phone ring before it was answered? _____

Were you placed on hold? _____ For how long? _____

APPENDIX 4

Please use this sheet to assign tasks that are important to maintain the image of your practice.

Job	Assigned to:	Frequency:	Initials:
Ex: Empty garbage in gym area	Mary	Twice daily	(Mary's initials)

APPENDIX 5

Questions for Patient Liaison (PL)

Directions: These are questions that potential patients may call and ask when they are shopping for a Physical Therapy practice to attend. Go through these questions with the PL and add any of your own. The purpose of this exercise is to train the PL to be able to answer any questions regarding the Operations of your clinic.

1) What are your hours of operation?

2) How long does a session last?

3) What insurances do you accept?

4) How much does it cost to come to Physical Therapy if I don't have insurance?

5) Do I need a doctor's prescription?

6) What will the Physical Therapist do during my session?

7) How difficult is it to get an appointment?

8) My doctor wants me to come for a month, will I always be with the same Physical Therapist?

9) I want to be with your best clinician. Who is the best Physical Therapist there? *Trick question! You should always support all clinicians equally.*

10) Do you have a traction machine?

11) Can I get put on the traction machine?

12) What kind of machines do you have?

13) How much of this will my insurance cover?

14) How many times does my insurance let me come in?

15) Do you treat (insert condition)? *Your PL needs a list and basic understanding of every condition you treat!*

APPENDIX 6

Weekly Responsibilities of the PL.

Day	Task
Monday	1. Go through the patient charts of all active patients. Update status and call patients if they are not on the schedule and should be. 2. Call all Patients scheduled for an Initial Evaluation on Tuesday or Wednesday and remind them of their appointment.
Tuesday	1. Make Internal Marketing calls to patients who are 2-3 weeks into treatment. 2. Set aside Thank You Postcards or Surveys for Physical Therapists to distribute to the aforementioned patients.
Wednesday	1. Set up Referral Rewards for Internal Referrals/friends and family referrals. 2. Call all patients who are scheduled for Initial Evaluations on Thursday or Friday.
Thursday	1. Make calls to Discharged patients. Keep track of Reactivations.
Friday	1. Call all patients who are scheduled for Initial Evaluations on Saturday and Monday to remind them of their appointments.

APPENDIX 7

Marketing survey

Patient Name: _____ Date: _____

How did you hear about our practice? Please check off all possible sources.

☐ My Doctor told me to come here. Doctor's name: _____

☐ My Doctor's nurse or receptionist told me to come here.

☐ A friend or family member recommended you. Friend's Name: _____

☐ I saw your ad in the paper.

☐ I saw your name in the Yellow Pages.

☐ I found you through my insurance.

☐ I found you online.

☐ I received a mailing from you.

☐ I saw your flyer in town.

☐ I saw your sign outside.

☐ Other:_____

Thank you!

<u>APPENDIX 8</u>

New Patient Packets should include the following three pieces of information:

1. **Welcome Letter:** A letter written to patients from the owner of the practice.

Dear Friend,

Welcome to Best Physical Therapy! Thank you for choosing us. I hope we can help you feel your best. If you ever have any concerns during the course of your Physical Therapy treatment, please don't hesitate to contact me. I am happy to help.

Yours in health,

Terri Korenstein
Owner, Best Physical Therapy
Personal Cell: 201-694-7738

2. **Things I Treat**: A list of the conditions you treat and services you provide. This can be in list form or appear on brochures or other marketing materials.

Best Physical Therapy provides treatment for so many conditions. We are happy to help you and your friends and family recover and feel your very best. We can help with:

- ACL tears and surgery
- Ankle injuries
- Back Pain
- Balance problems
- Dancer's hip
- Frequent falling
- Golfer's elbow
- Herniated discs
- Hip pain
- Knee pain
- Labral tears
- Meniscal tears and surgery
- Multiple Sclerosis
- Neck Pain
- Osteoarthritis
- Parkinson's Disease
- Post surgical rehabilitation
- Post surgical rotator cuff repairs
- Rheumatoid Arthritis
- Rotator cuff tears
- Sciatic nerve pain
- Shoulder pain

- Tennis elbow
- Total joint replacements

Services:

- Spine Clinic
- Osteoporosis Exercise Class
- Fit after 50 Exercise Class
- Pregnancy and Back Pain Class
- Pilates

3. **Greatest Gift:** This letter politely asks for referrals from your current patients.

Dear friends,

The greatest gift we can receive is the referral of your friends and family. We love what we do and take great pride in helping you feel your best. When you grant us the privilege of working with those you care about, we are truly grateful. Thank you for your support.

From all of us at Best Physical Therapy

APPENDIX 9

Sample Thank You Notes to Doctors

1. Dear Doctor Smith,

    Thank you for referring Jane Doe to Best Physical Therapy. I will do everything I can to help her feel her best. I appreciate your support and confidence in our practice.

2. Dear Doctor Smith,

    Thank you for the support you have shown Best Physical Therapy over years. We are celebrating our 5th Anniversary soon and I know that your support was integral to the success of our practice.

3. Dear Doctor Smith,

    Thank you for your support of Best Physical Therapy. I love what I do and take great pride in helping your patients feel their best. If you have any feedback that would improve our treatment of your patients, please don't hesitate to contact us.

4. Dear Doctor Smith,

    It's only been a few short months since we opened our doors and we are thrilled to be a part of this medical community. Recently, I treated your patient, Jane Doe, for knee pain. Thank you for your support. Please let us know if you ever have any concerns or feedback for us.

5. Dear Doctor Smith,

    Thank you for the opportunity to assist your patients in their Physical Therapy needs. I appreciate your support.

6. Dear Doctor Smith,

    I have heard such positive things about you from your patients. I hope to have the opportunity to meet you in the near future. Thank you for your support of Best Physical Therapy.

7. Dear Doctor Smith,

    Thank you for the kind referral of Jane Doe to Best Physical Therapy. We appreciate your support and the privilege of working with your patients.

8. Dear Doctor Smith,

   Thank you for referring your patient, Jane Doe, to Best Physical Therapy after undergoing cervical fusion surgery with you. We will do everything possible to help her heal, according to your prescription blueprint.

9. Dear Doctor Smith,

   Thank you for supporting Best Physical Therapy over the years. We take referrals very seriously and thank you for trusting us with your patients. We will always do everything possible to restore your patients to the best of health.

10. Dear Doctor Smith,

    Best Physical Therapy values the referrals you have given us since opening our doors in 2010. Thank you for your ongoing support and confidence in our practice.

APPENDIX 10

Trend Report: This report is created from the data collected in your Marketing Surveys.

Dates: _____

List the dates covered in your Marketing Surveys. Example, June 1- June 5.

These headings directly fr[o] your Mark[eting] Surveys.

Patient's Name	Doctor	Nurse	Friend/ Family	News-Paper	Yellow Pages	Insurance	Website	Mailer	Flyers	Sign
Doe, Jane			X							
Williams, Peter	X									
Jones, Katie						X				
James, Sandy			X							
Total: 4 patients	1		2			1				

Use the patients named in your marketing surveys and check off their responses in this chart.

Add up the total in each column.

Ask patients to fill out a postcard that you will send to their doctor.

Thank You For Sending Me to Best Physical Therapy!

Dear Dr. _____

Ask the patient if they would like to send a postcard like this to their doctor. Let them know that you will address it for them and mail it out.

From:
TCG
PO Box 122
Little Falls, NJ 07424

Place your stamp here

To:

_____
_____
_____

APPENDIX 12

At the midpoint of Physical Therapy, please ask patients to fill out this survey and let them know that the contents will be shared with their referring physician.

- - - - - - - - - -

Dear Friend,

Please take a moment to let us know how you are feeling. Please consider the following:

- How are you feeling?
- Are you pleased with the care you are receiving?
- What can you do now that you couldn't do before?

Please share these thoughts and any other comments in the space below.

Signature: _____Date: _____

APPENDIX 13

Sample script for PL's Internal Marketing Phone Calls.

These calls should be made around the patient's 2 week mark in Physical Therapy.

- - - - - - - - - -

Hello! My name is Terri and I am a Patient Liaison at Best Physical Therapy. This is just a courtesy call. I typically help patients setup their first appointment and help with any needs throughout the duration of your time with us. You have been coming to Physical Therapy for a few weeks now.

Are you pleased with the care you are receiving?

If you ever have any concerns, questions, or feedback, please don't hesitate to contact me. We want your experience with us to be a positive one and we want you to feel your best. You can reach me at (phone number).

Take care!

APPENDIX 14

When you are going through your written marketing materials, such as brochures and flyers, consider the following:

Appearance
- Are your materials written on quality paper stock or is the material flimsy?
- Are there creases and tears on your marketing materials?
- Are your materials visually appealing? Do they have a lot of blocks of text?
- Should you break up text with pictures or changes to the font and characters?

Content
- Are there any typos?
- Is the grammar current?
- Is it easy to tell what the message is?
- Have you accurately represented your practice in these materials?

Website Design Basics
- Is the location and phone number of your practice visible on the home page without having to scroll?
- Most people focus on the top left corner of a new website. Are you maximizing this space with important information?
- It has been shown that it is more visually appealing to keep some empty or white space on a website. Is your website full or do you have some breaks of empty space?
- Have you gone overboard with texture, graphics, and different fonts? Is your website easy to read?
- Most people scan websites for useful information. Do you have a lot of long paragraphs or do you provide snippets of information in short paragraphs?
- Does your website take a long time to load?

APPENDIX 15

It's a good idea to offer a Freebie of some kind to get prospective patients in your doors. Here is an example of a Complimentary Consultation Coupon that can be modified for your practice.

The bearer of this coupon is entitled to:

A Complimentary Consultation
at Best Physical Therapy. Valued at $99!

PO Box 122, Little Falls, NJ 07424
Please call to schedule.

Signature: _____
Not redeemable for cash

www.HealthMarketingStrategies.com

APPENDIX 16

Name: _____ Job Title: _____

Please list your job responsibilities:

-
-
-
-
-
-
-
-
-
-
-
-
-
-

I understand the tasks expected of me in this job.

Signature: _____ Date: _____

Supervisor's Signature: _____ Date: _____